100 RADIO HOOKUPS

by

MAURICE L. MUHLEMAN

PRICE
25¢

2nd EDITION REVISED

THE E.I. CO.
NEW YORK CITY

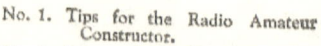
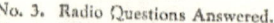

To order additional copies of this book

go to www.CreateSpace.com

THE EXPERIMENTER'S LIBRARY, No. 7

100 Radio Hook-ups

By
MAURICE L. MUHLEMAN
on the Staff of RADIO NEWS

SECOND EDITION, REVISED

Published by THE E. I. COMPANY
233 Fulton Street
New York City

Copyright by
THE E. I. COMPANY
1924

1st Edition, May, 1923
1st Reprint, June, 1923
2nd Reprint, July, 1923
3rd Reprint, August, 1923
4th Reprint, October, 1923
5th Reprint, November, 1923
6th Reprint, January, 1924
7th Reprint, February, 1924
8th Reprint, March, 1924
9th Reprint, May, 1924

FOREWORD

BEFORE entering upon the study of the hook-ups described in the pages of this book, the reader would do well to acquaint himself somewhat with the various instruments used in the reception of radio telegraph and telephone messages.

The first instrument that the layman will, in all probability, lay his hands on is some sort of a tuning coil. This is nothing more or less than many turns of wire wound in an even layer on an insulating core and equipped with some method of varying the number of active turns. This method may be either a slider or a number of leads brought off from the coil to an equal number of switch points placed in an arc of a circle and so arranged that a switch-arm will make contact with one or another of them. The purpose of this coil is to boost the natural wavelength of the aerial to that sent out by the transmitting station. Other instruments that perform the same purpose are Honeycomb and Spider-web coils, variometers, vario-couplers and loose-couplers. The variometer is a very finely adjustable inductance coil. Its action will be found described in any up-to-date radio text book. The author does not consider this book to be the place for full discussions of the action of the various instruments so it will not be dwelt upon.

If two coils of wire are placed near one another, and an alternating current run through one, a current will be induced in the other. On this principle the action of the variometer, vario-coupler and loose-coupler depend. In the former the two coils are connected together, while in the latter two they are "coupled" only by induction. In radio work it has been found that the inductively coupled tuners give greater selectivity while those that are "coupled" by wires, or conductively coupled, give the loudest signals. If the amateur is in a district where there are many broadcasting stations in the vicinity he will do well to choose a set employing the former method of tuning. Honeycomb and Spiderweb coils may also be arranged so as to permit this use, and several diagrams of them used in such a way are shown in the following pages.

Variable condensers also play an important part in the tuning of a radio set. They also vary the wavelength to which the receiving set will respond. They are almost always used in connection with a coil of some kind. When used with Spider-web or Honeycomb coils they are connected either in series with the coil or across it as is found best. In the former position, they reduce the wavelength of the coil, in effect, while in the latter position, they increase it.

These condensers usually consist of several semi-circular plates of aluminum so arranged that alternate plates revolve, inter-leaving with each other but not touching. They may take other forms but the majority of them are as described above. When used in a radio receiving set, the effect of variable condensers is to make the tuning of the set sharper, that is, make the circuit capable of eliminating one station when it is desired to listen to another operating on nearly the same wave. They also enable the receiving operator to tune between the smallest number of turns which his tuner controls.

4

The detectors to be considered by the layman in selecting a set may be roughly placed in two classes; the crystal and the audion or vacuum tube. The first is the cheapest and will bring in the clearest signals, but only over a comparatively short range. The latter is very sensitive to signals from stations at a distance and will bring them in much louder than a crystal set. The greatest judge here is the pocketbook of the reader. Whatever you buy, however, be sure that it is of the best, as cheap instruments are sometimes worse than useless.

A word regarding regeneration will not be amiss here. When a vacuum tube or audion is used in a receiving circuit, there is set up, in the plate circuit of the tube, a current which will actuate the phones. If now this current is fed back into the grid circuit, a great increase in volume will be found. This same effect, because of the characteristics of the audion, may be obtained by tuning the plate circuit of the tube to the same value as the grid circuit. This method of producing regeneration is usually known as the tuned plate circuit. Many variations of the above two methods are in use today with greater or less satisfaction and some of them are shown in the following pages. The super-regenerative tuner designed by Major Armstrong, the inventor of the regenerative circuit, works on somewhat the same lines.

SYMBOLS USED IN HOOK-UPS

	AERIAL			CHOKE COIL	
	AERIAL (LOOP)			COIL	
	ALTERNATOR			COIL (HONEYCOMB)	
	AMMETER			COIL (SPIDERWEB)	
	ARC			COIL (TUNING) (VARIABLE INDUCTANCE)	
	BATTERY "A"			CONDENSER (FIXED)	
	BATTERY "B"			CONDENSER (VARIABLE)	
	BUZZER			CONNECTION	

The left shows a picture of the apparatus while the right shows the symbols used in all radio hook-ups.

6

SYMBOLS USED IN HOOK-UPS

	DETECTOR (CRYSTAL)			LOOSE COUPLER COUPLED COILS WITH VARIABLE COUPLING	
	DYNAMO OR MOTOR			NO CONNECTION	
	GAP (SPARK)			PLUG	
	GAP (QUENCHED)			POTENTIO-METER	
	GROUND			RECEIVER (TELEPHONE)	
	GRID LEAK			RESISTANCE (VARIABLE) FILAMENT RHEOSTAT	
	JACK			RESISTANCE	
	KEY			SWITCH	

The left shows a picture of the apparatus while the right shows the symbol used in all radio hook-ups.

SYMBOLS USED IN HOOK-UPS

	TRANS-FORMER (AUDIO FREQUENCY)			VACUUM TUBE	
	TRANS-FORMER (RADIO FREQUENCY)			VARIOMETER	
	TRANS-MITTER				
	VOLTMETER	—Ⓥ—		VARIO-COUPLER	

The left shows a picture of the apparatus while the right shows the symbol used in all radio hook-ups.

HOW TO CONNECT RADIO INSTRUMENTS

THE three illustrations, Figure A, Figure B, and Figure C, show perspective views of how radio instruments are usually wired up.

When making such connections, it is absolutely important that all connections be very clean and bright, and wherever wires are connected to themselves, outside of the instruments, they should be soldered.

All connections must be just as short as it is possible to make them—superfluous lengths of wire only make for poor results and low efficiency.

The three pictures, A, B and C, have been shown purposely to demonstrate how difficult it is to study a circuit in this way. For that reason, the usual hook-ups are used.

For instance, diagram A is the same circuit as hook-up No. 2 shown on page 14. You can readily see that the circuit can be traced better once you have grasped the symbols in Figure 2, than if you traced the circuit in diagram A.

Diagram B refers to hook-up No. 10, shown on page 15, and here again you will see that it is easier to follow the circuit of the hook-up than the picture.

Diagram C shows a picture layout of the instruments assembled in hook-up No. 69, illustrated on page 37.

When it becomes necessary to actually make your connections, to hook up any particular outfit, it may be well to make a picture diagram, similar to diagrams A, B, and C, and lay out the connecting wires in different col-

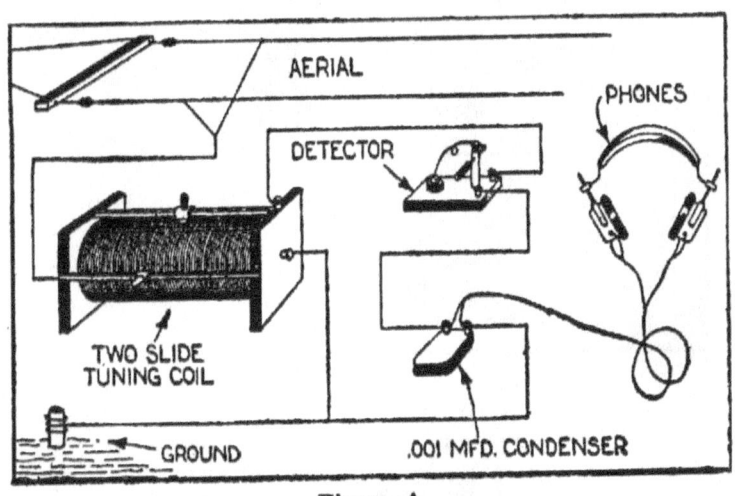

Figure A

ored pencil, such as black, red, blue, yellow, before you start your wiring.

Then, as you proceed with your work, cross off each wire as the connection has been made. You will then not have any difficulty in making either wrong connections or leaving out a certain connection.

Proceed slowly and cautiously when wiring. Never hurry the job. You will get better results in the end.

HOW TO READ A HOOK-UP

THE intelligent reading of a hook-up appears rather difficult at first glance. This naturally is due to the maze of wiggly lines that mean nothing to the person not acquainted with radio symbols. The use of symbols is the most rapid and practical method

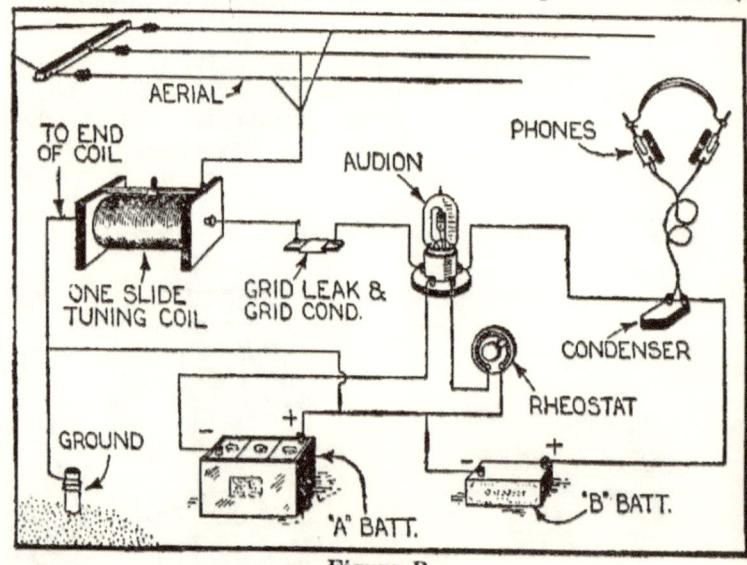

AERIAL

TO END OF COIL

PHONES

AUDION

ONE SLIDE TUNING COIL

GRID LEAK & GRID COND.

CONDENSER

RHEOSTAT

GROUND

"A" BATT.

"B" BATT.

Figure B

of illustrating a complete receiving set and its connections. This method will never give way to the long drawn out means of showing each instrument in perspective form. It is necessary for the beginner though, to have perspective illustrations of each piece of apparatus so that he can compare it to its line symbol. For this reason the author has included a list of the components that all or in part make up a receiving set. A number of other symbols are included for general information. These are seen to be understandable drawings of the instruments with the correct symbol alongside each one. Since the symbols are more or less descriptive of the instrument it designates, it should not take the reader long before he can readily determine the meaning of each symbol. Let us take for an example the hook-up of Fig. 1. It composes in the drawing of an aerial, a ground, a one slide tuning coil, a crystal detector, a fixed condenser and head phones. If any of these symbols are foreign in their appearance, their nature can easily be found by

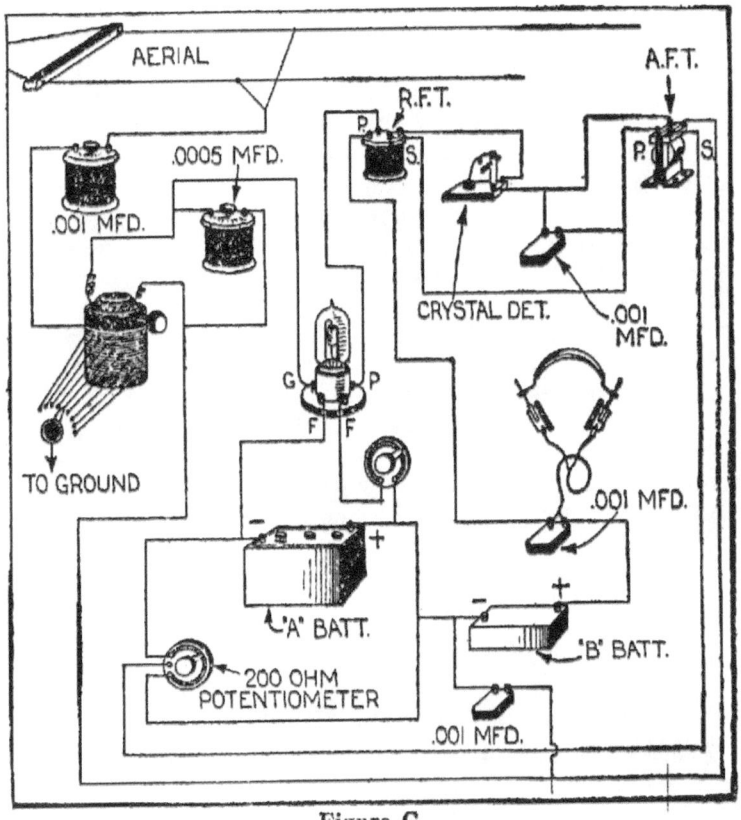

Figure C

referring to the list here given. The straight lines issuing from each side of the symbols are the connecting wires, which in practice, lead from the binding post of one instrument to the binding post of another. In some cases there may be another wire, or branch so to speak, from between a connection, leading to some other terminal. This is usually termed a common connection (being connected to two or more instruments). Many of these are found in vacuum tube hook-ups.

How to Connect Up the Instruments

Many mistakes are made in the wiring of a receiving set. These in most cases are due to the complication of wiring, the following of which requires more than usual care. It is suggested that the circuit to be used be traced with pencil on to a sheet of paper. Starting from the most convenient point in the hook-up, connect one wire at a time, and for each wire connected, go over it with ink on the diagram and give it a number. It becomes an easy matter then to check up on all connections made. The going over of each connection after the set has been completely wired is recommended. A mistake may cause considerable trouble later.

Don't attempt to use a complicated hook-up at first. Be content with a small outfit until you master it. After you look over the diagrams in this book, select the instruments so that those in your first set will be of use when you branch out to a bigger lay-out.

The best arrangement for the various instruments composing the receiving set is exactly as they are positioned in the diagrams. This arrangement allows for the shortest possible leads from one instrument to another, which is a decided advantage.

Concerning Tubes

There are a number of different types of vacuum tubes on the market today, some of which are designed for a definite function, and others which are well suited for a variety of functions. Poor results are often due to the use of the wrong type of tube.

For detectors and regenerators any one of the following tubes can be used. Radiotron U.V. 200, Cunningham C 300, Western Electric VT 1 (J tube), or Westinghouse WD-11. The Radiotron U.V. 201, Cunningham C 301 or the Western Electric VT 2 function well as detectors and regenerators if a "B" battery of about 40 volts is used instead of the usual 22½ volts.

For audio frequency amplifiers the Radiotron UV 201, UV 201-A, Cunningham C 301, C 301-A, and the Western Electric VT 1, VT 2 and 216-A are all well suited. The Westinghouse WD-11 can be used also, if excessively high "B" battery voltage is not employed. About 45 volts "B" is correct for this tube.

For radio frequency amplifiers the VT 1, UV 201, C 301 and the WD-11 are probably the best tubes to use. The UV 201-A, C 301-A and VT 2 can also be used but they are not quite as efficient as the others for this purpose.

When audio frequency power amplification is desired use either a Radiotron UV 201-A, C 301-A, VT 2 or a 216-A. These tubes have a higher amplification constant than the others and much higher "B" battery voltages can be used.

All of the above mentioned tubes, except the WD-11, require a six volt storage battery to supply the necessary current for lighting the filaments. The WD-11, however, requires only a 1½ volt dry cell.

CRYSTAL HOOK-UPS

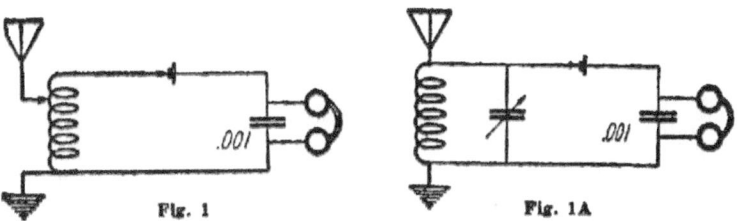

Fig. 1   Fig. 1A

Fig. 1. A crystal circuit employing a one slide tuning coil which will give fair results for local receptions. Fig. 1A is an adaption of this circuit to a coil which is not tapped or provided with a slider. In this case the wave length range is controlled by a variable condenser.

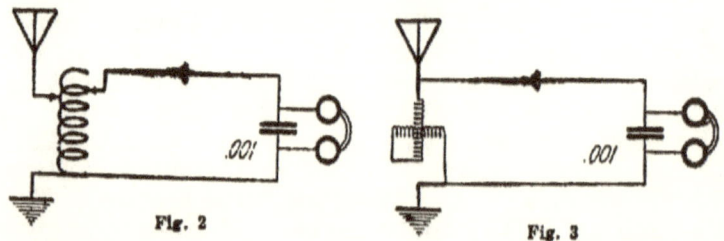

Fig. 2   Fig. 3

Fig. 2. By using a two slide tuning coil, greater selectivity will be obtained due to the closer variation of the number of turns of wire used.

Fig. 3. A circuit having only one control may be used, a variometer doing all the tuning. This circuit is fairly selective and has the added advantage that no sliding contacts are used.

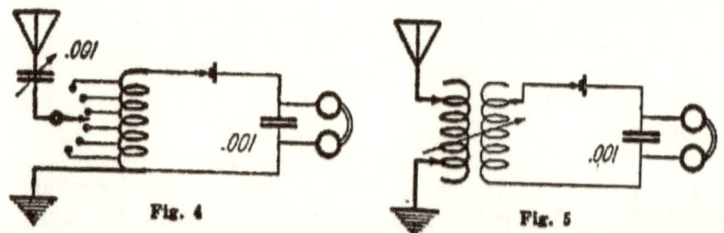

Fig. 4   Fig. 5

Fig. 4. A tuning coil that has taps instead of a slider may be used by connecting a variable condenser in the circuit as shown. The latter accomplishes quite sharp tuning.

Fig. 5. A still more selective circuit may be obtained by using a loose coupler. This, however, will not yield as loud signals as the former circuit.

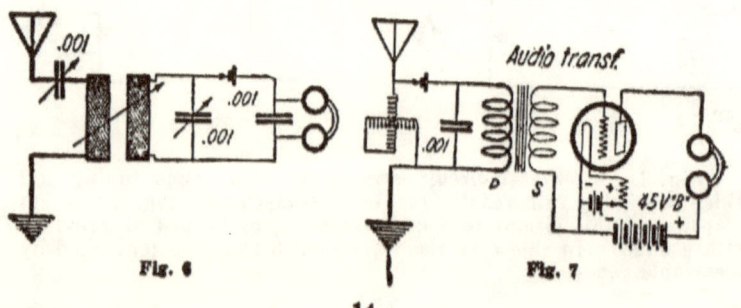

Fig. 6   Fig. 7

14

Fig. 6. Honeycomb coils may be used in place of a loose coupler with practically the same results.

Fig. 7. By amplifying the signals detected by a crystal detector with an audion tube much louder signals will be obtained. This arrangement can be adapted to all the previous circuits shown.

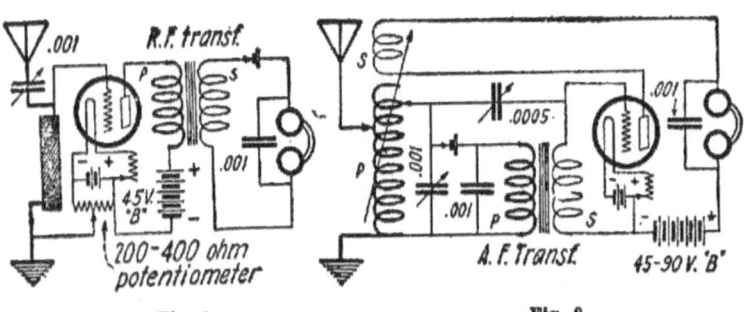

Fig. 8   Fig. 9

Fig. 8. Incoming signals may be amplified before they are detected as shown. This circuit gives reception over much greater distances than with the crystal detector alone, or with Fig. 7.

Fig. 9. A regenerative amplifying circuit using a crystal detector and one stage of audio frequency amplification is shown. The two coils may be the primary and secondary of the vario coupler.

PLAIN VACUUM TUBE HOOK-UPS

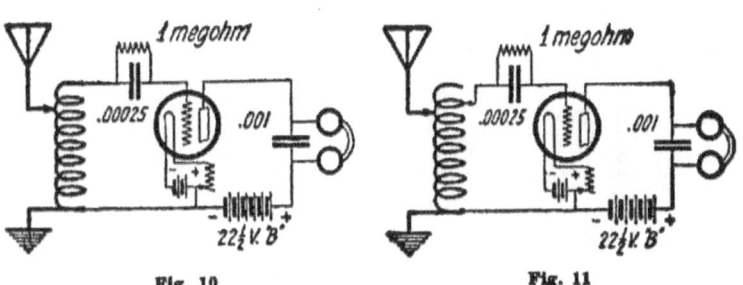

Fig. 10   Fig. 11

Fig. 10. A one slide tuning coil may be used in connection with the audion tube with fair results.

Fig. 11. Greater selectivity than that obtained with Fig. 10 may be had by employing a two slide tuner.

15

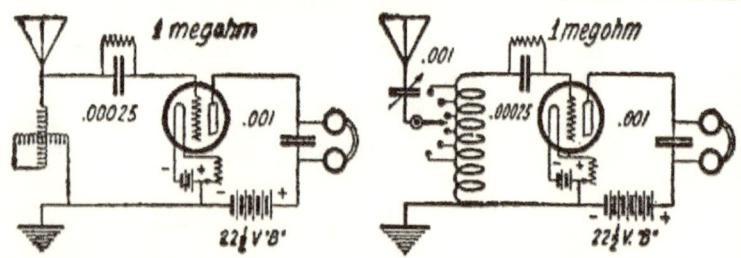

Fig. 12

Fig. 13

Fig. 12. A one control circuit employing a variometer will give good results.

Fig. 13. A fairly selective circuit employing a tapped coil may be connected as shown.

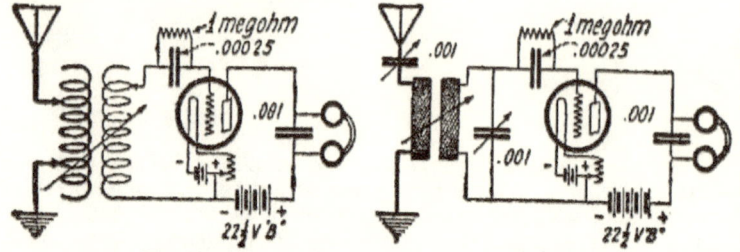

Fig. 14

Fig. 15

Fig. 14. Much greater selectivity may be obtained with an audion detector by employing a loose coupler.

Fig. 15. Two Honeycomb coils make a very good tuner in which the tuning is accomplished by means of variable condensers.

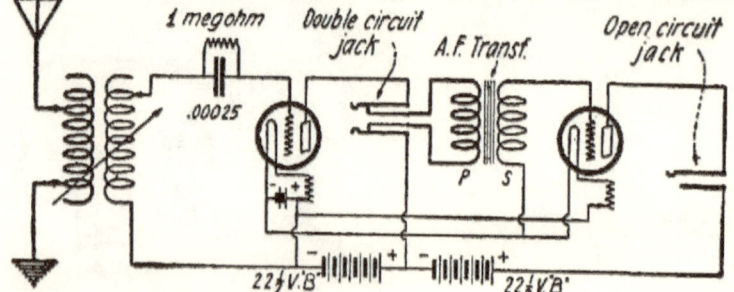

Fig. 16

Fig. 16. The connections of a detector and one stage audio frequency amplifier are shown. Jacks are used so that the head phones may be plugged into the detector circuit or into the amplifier as desired.

16

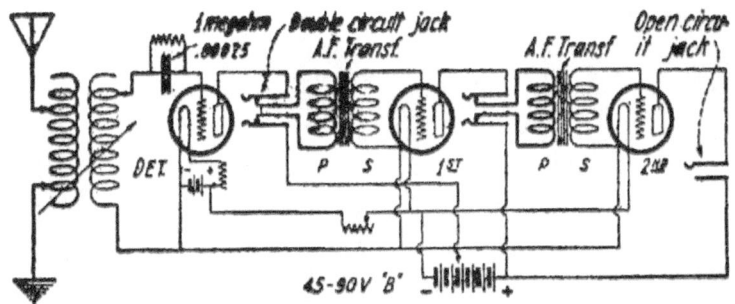

Fig. 17

Fig. 17. A standard vacuum tube set employing two stages of audio frequency amplification. Jacks are also used here for convenience.

REGENERATIVE VACUUM TUBE HOOK-UPS

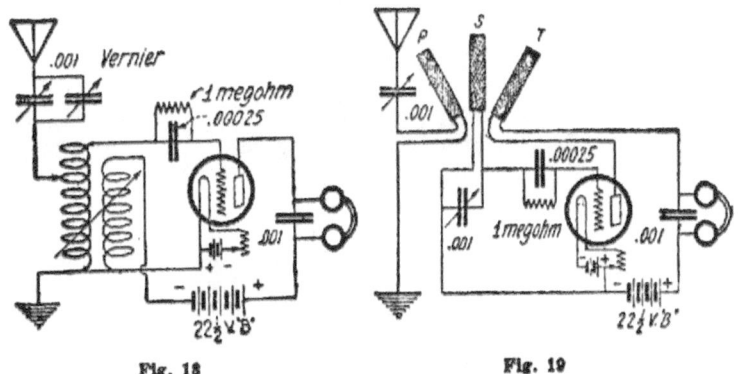

Fig. 18   Fig. 19

Fig. 18. The simplest form of a single circuit regenerative tuner is shown. The two coils may be the primary and secondary of a vario coupler.

Fig. 19. Three Honeycomb coils connected as shown form a very selective and stable circuit. Tuning is, of course, accomplished by the variable condensers.

17

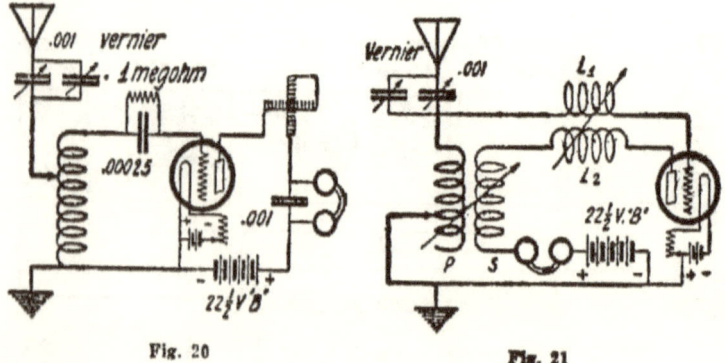

Fig. 20   Fig. 21

Fig. 20. The tuned plate method of securing regeneration is shown. Sharp tuning is accomplished by means of the vernier variable condenser.

Fig. 21. An excellent regenerative circuit employ two Spider-web coils marked L-1 & L-2 and a vario-coupler indicated by P. & S. This circuit is capable of very fine adjustment.

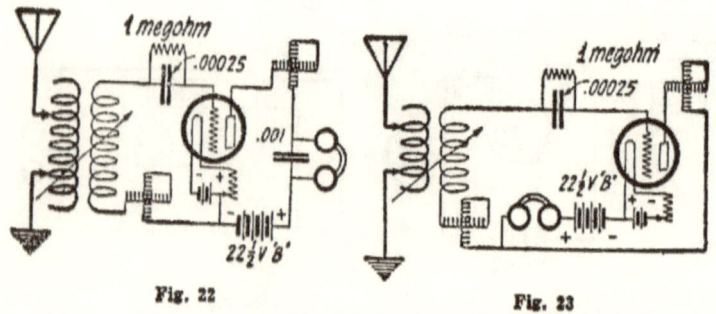

Fig. 22   Fig. 23

Fig. 22. A standard three circuit regenerative receiver employing a vario-coupler and two variometers for tuning and regeneration. This is one of the most sensitive and selective sets in common use today.

Fig. 23. This is an improvement over the circuit shown in Fig. 22 since the phones and "B" batteries are taken out of the main oscillating circuit. This arrangement, however, makes the set much more critical in adjustment.

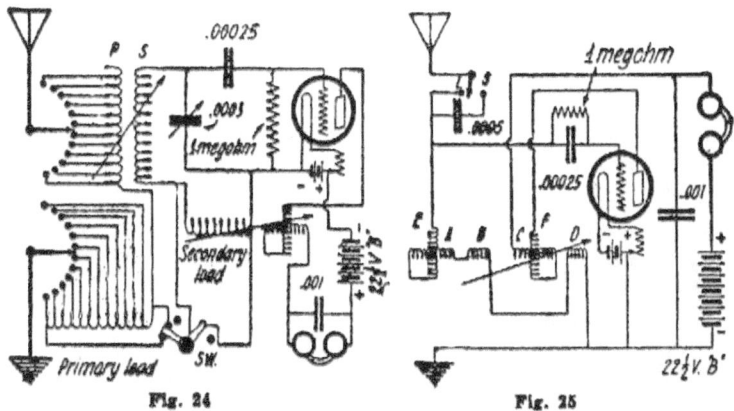

Fig. 24   Fig. 25

Fig. 24. This is the circuit of the Paragon RA-10. A "load-ing coil" is inserted in both the primary and secondary circuits so that higher wave-lengths can be reached. These coils are cut in and out of the circuit by switch "SW".

Fig. 25. The hook-up of the Aerolia Sr. Coils A, B, C and D are all wound on the same tube while coils E and F are wound on smaller tubes and rotate within the others. When the aerial switch is on contact "S" shorter waves can be received than when on contact "L".

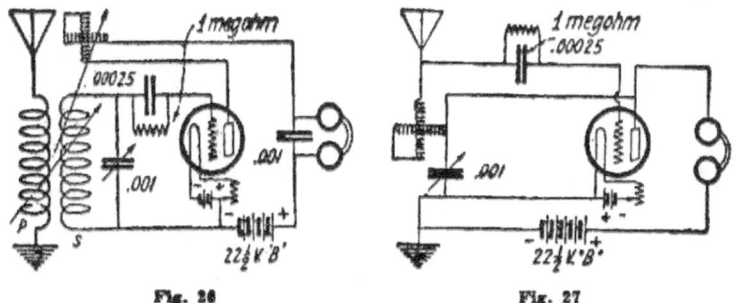

Fig. 26   Fig. 27

Fig. 26. A very good regenerative receiver with but two controls, i. e., the variometer and the variable condenser. Coils "P" and "S" are wound on the same tube with about ½ in. be-tween them. The variometer should be placed at the end of this tube. Coil "P" should be "stagger" wound.

Fig. 27. Another circuit with but two controls. This is an adaptation of the Colpitts oscillator circuit and has proven to be a very sensitive regenerative receiver.

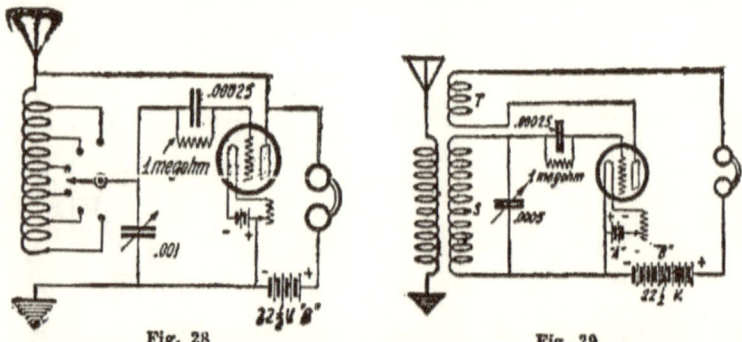

Fig. 28   Fig. 29

Fig. 28. The De Forest single coil Ultra-audion circuit. A single layer tapped coil is employed rather than a honeycomb coil, so that a less critical means of wavelength variation can be had. The Ultra-audion circuit, although very sensitive, is rather unstable in operation. Better control can be had by the use of a variable grid condenser having a maximum capacity of .0005 mfd.

Fig. 29. A regenerative circuit employing an untuned primary coil. Coils S and T are the stator and rotor of a vario-coupler respectively. The primary P consists of 10 turns of No. 18 D.C.C. wire wound directly over and in the center of the stator coil of the variocoupler. This is a more selective arrangement than the single circuit tickler feed back regenerative receiver.

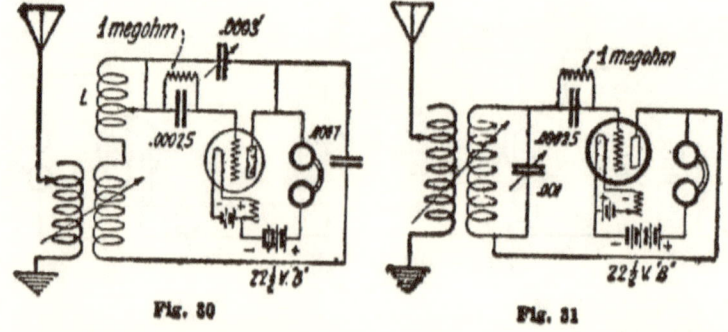

Fig. 30   Fig. 31

Fig. 30. Here is one type of capacity feedback regenerative receiver which has proven very satisfactory for long distance work. A small variocoupler is used with the extra tapped coil "L".

Fig. 31. Still another type of capacity regenerator. It is really a simplified form of the one shown in Fig. 30. A standard variocoupler is employed.

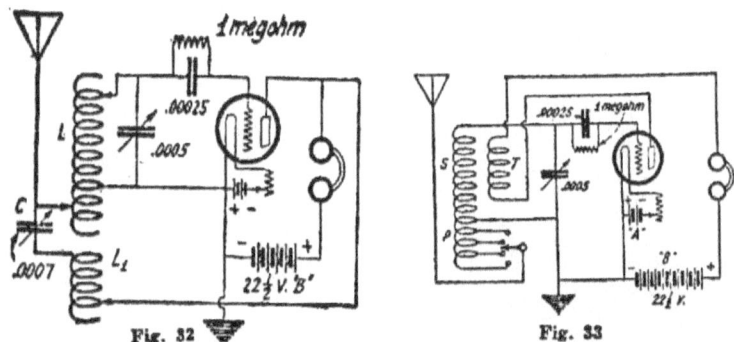

Fig. 32   Fig. 33

Fig. 32. The famous Reinartz circuit. Coils L and L-1 are wound in "spider-web" fashion on the same form, coil L-1 acting as a tickler. There is also a capacity feedback through the condenser C.

Fig. 33. This is the recently exploited Haynes circuit and is identical to a single circuit tickler feedback regenerative receiver except for the conductively coupled primary circuit P. T is the rotor coil of a variocoupler and S-P the stator coil. The first seven single turn taps of the variocoupler are run to switch points, and the eighth to the variable condenser, as shown. This circuit has a fair degree of selectivity.

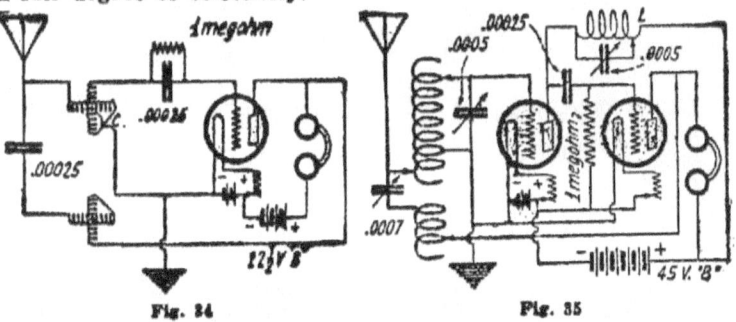

Fig. 34   Fig. 35

Fig. 34. An adaption of the Reinartz circuit wherein the control is accomplished by two variometers. It is a combination capacity feedback and tuned plate circuit. The center tap "C" is taken from between the two stationary coils of the variometer capacity and inductive feedback circuit. The center tap "C" is taken from between the two stationary coils of the variometer

Fig. 35. This shows the original Reinartz circuit incorporated with one stage of radio frequency amplification, this being of the tuned impedance type. 35 turns of wire on a 3 in. tube will suffice for coil "L".

21

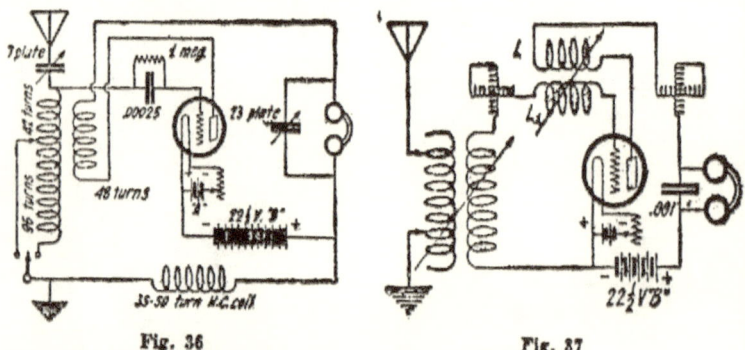

Fig. 36   Fig. 37

Fig. 36. Here is the Long 45 circuit. It is of the single circuit tickler feedback type with an extra coil connected in the plate-grid circuit, this presumably, to aid regeneration. The 23 plate variable condenser is employed to by-pass the radio frequency currents around the headphones.

Fig. 37. A three circuit regenerative receiver. L and L-1 are two small Spider-web coils of about 20 turns each. Coil L is movable in relation to L-1. The variometer assists in controlling the regeneration.

COMBINATION HOOK-UPS

Standard Amplifying Transformers Will Give Good Results in All the Following Circuits

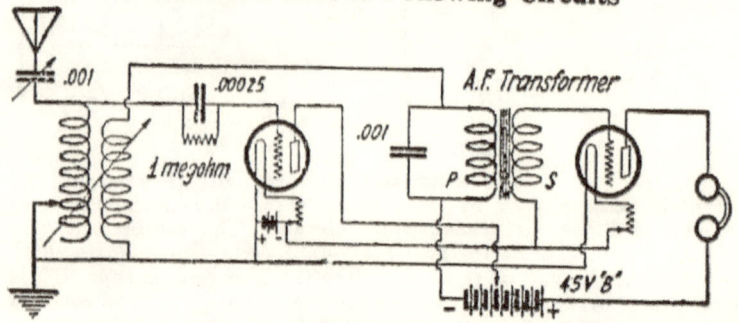

Fig. 38

Fig. 38. A single circuit regenerative receiver with a one stage audio frequency amplifier. A loud speaker can be used with this when receiving from stations in the near vicinity.

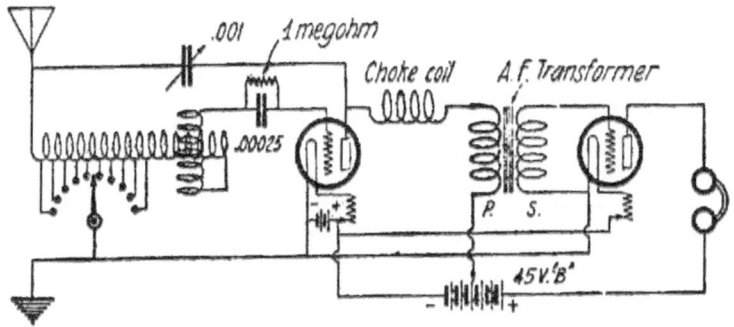

Fig. 39

Fig. 39. The improved Reinartz circuit with one stage of audio frequency amplification. It is advisable to insert a radio frequency choke coil in the circuit as shown. About 100 turns of No. 26 D.C.C. wire on a 3 in. tube will do for this.

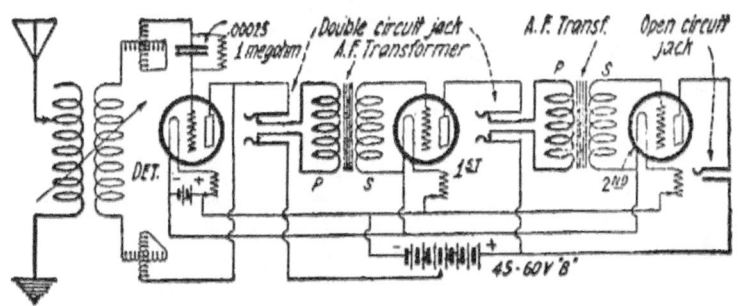

Fig. 40

Fig. 40. A three circuit regenerative receiver with the addition of a two stage audio frequency amplifier. Jacks are provided so that the phones can be plugged in on the detector, first stage or second stage.

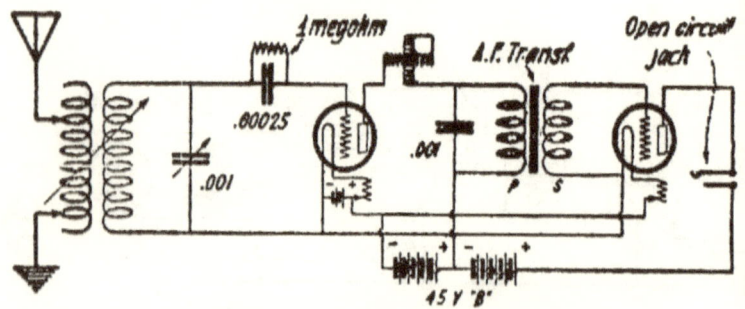

Fig. 41

Fig. 41. A tuned plate regenerative receiver employing a variocoupler, variable condenser and a variometer in connection with a one stage audio frequency amplifier. The jack shown can be replaced by two binding posts but it becomes handy when different phones or a loud speaker are often interchanged.

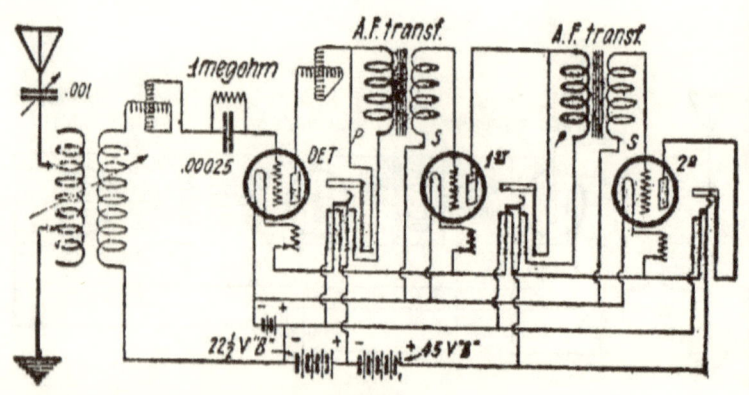

Fig. 42

Fig. 42. This is the same arrangement as shown in Fig. 40 except that filament control jacks are used. Plugging in on any of these jacks automatically lights the filaments of the tubes in use which are cut off when the plug is taken out.

AUDIO FREQUENCY AMPLIFIER UNITS

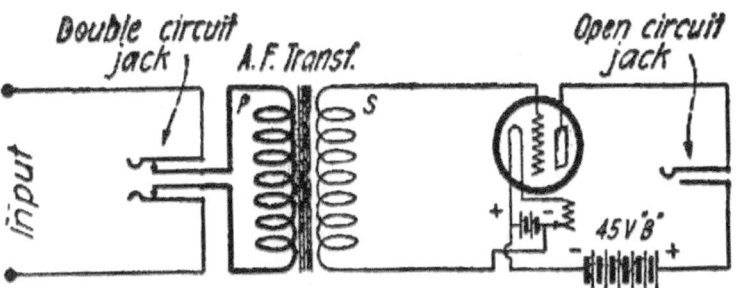

Fig. 43.

Fig. 43. This shows the connections for a one stage audio frequency amplifier when used as a separate unit. The "Input" posts connect directly to the "Phone" posts on the receiving set.

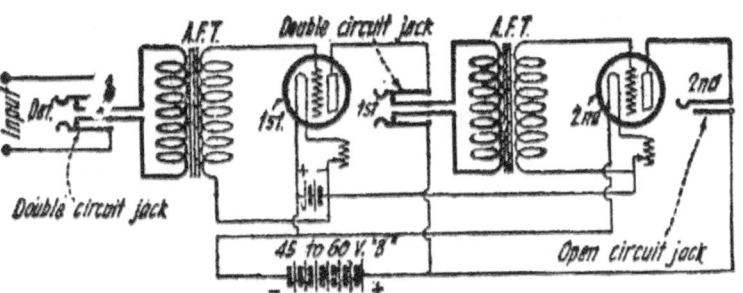

Fig. 44

Fig. 44. A two stage audio frequency amplifying unit with the necessary jacks and batteries.

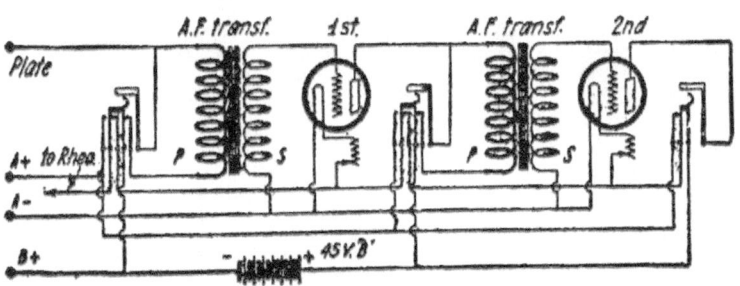

Fig. 45

Fig. 45. The same unit as shown in Fig. 44 except that filament control jacks are employed. The post marked "Plate" would go to the phone post on a receiving set which connects to the plate, while the post marked "B+" would go to the other phone post. The A+ and A— posts connect directly to the storage battery.

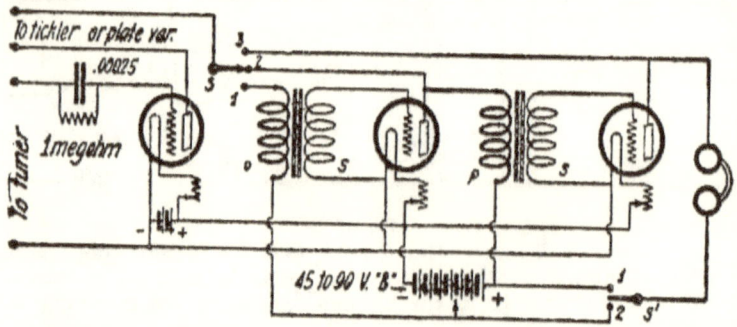

Fig. 46

Fig. 46. A two stage audio frequency amplifier wherein the successive stages are controlled by two small switches S and S-1. For detector alone, switch S is placed on contact 3 and switch S-1 on contact 2. For detector and one stage S is placed on 2 and S-1 on 1. For detector and two stages, S is placed on 1 and S-1 on 1.

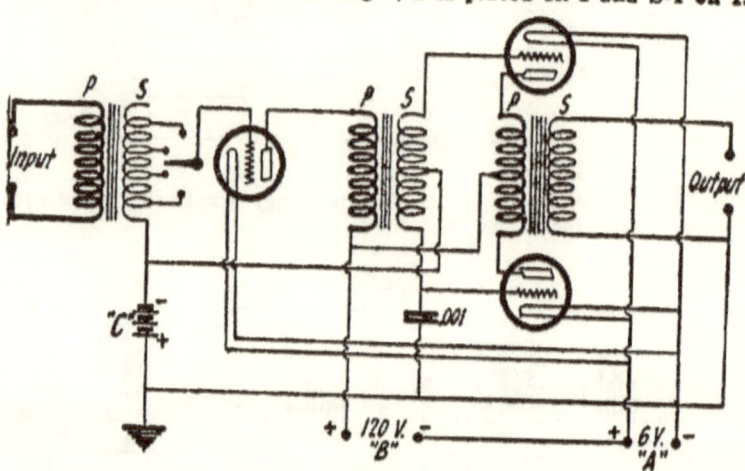

Fig. 47

Fig. 47. The Push-Pull Amplifier circuit as used in the Western Electric Loud Speaker. Greater amplification can be obtained with this arrangement.

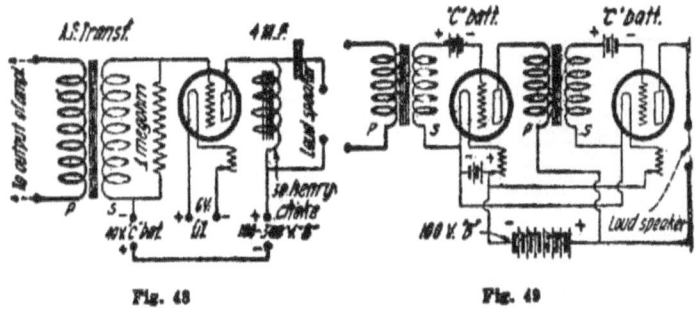

Fig. 48   **Fig. 49**

Fig. 48. A one stage audio frequency power amplifier which can be connected directly to the output terminals of the regular amplifier. The audio frequency transformer should be one that will stand the high voltage used without breaking down. The value of the grid leak can best be determined by experiment.

Fig. 49. Greater volume can be had from a standard audio frequency amplifier by the insertion of a "C" battery in the grid circuit of each amplifying tube as shown. Be sure that the negative side is connected to the grid.

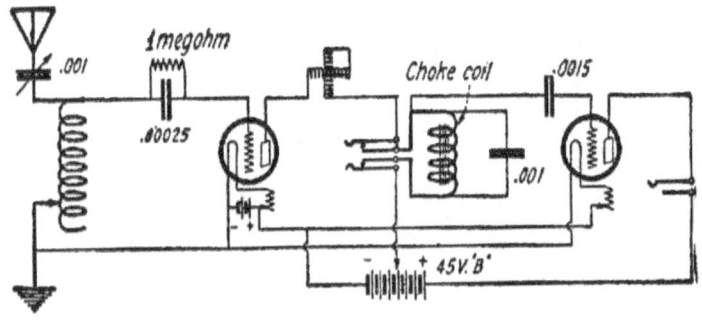

Fig. 50

Fig. 50. This shows a regenerative receiver connected to a one stage choke coil amplifier. A condenser must be placed in the grid circuit of the amplifying tube as shown.

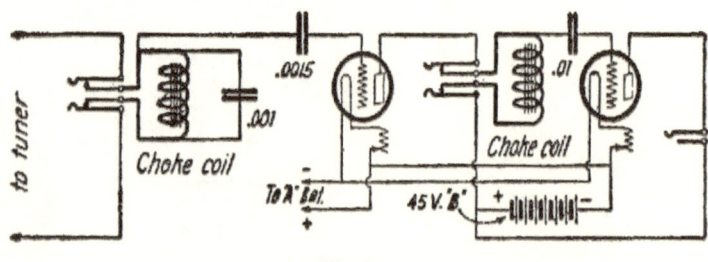

Fig. 51

Fig. 51. A two stage choke coil amplifying unit with jacks. "B" battery voltage up to 90 can be used if desired.

RADIO FREQUENCY AMPLIFIERS

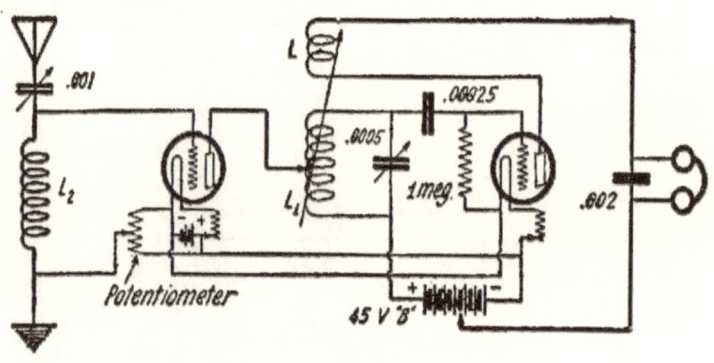

Fig. 52

Fig. 52. A simple single circuit vacuum tube receiver with a one stage tuned impedance radio frequency amplifier. Regeneration is obtained by the coil L which acts as a tickler. L-2 may be a 35, 50 or 75 turn Honeycomb coil. Coils L and L-1 can be those of a variocoupler. A potentiometer is required in this circuit and should have a resistance of from 200 to 400 ohms.

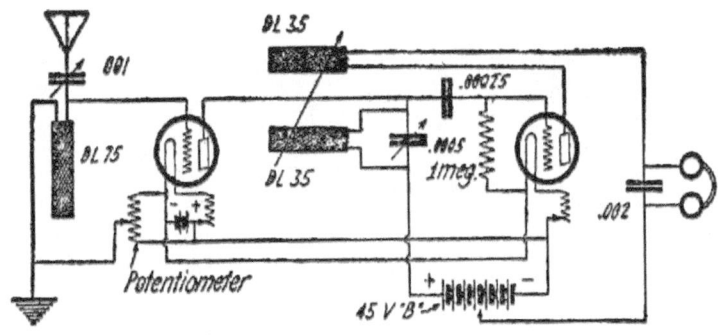

Fig. 53

Fig. 53. This is the same arrangement as that of Fig. 52 except that Honeycomb coils are used. The DL 75 coil should be at right angles to the other two, and separated therefrom.

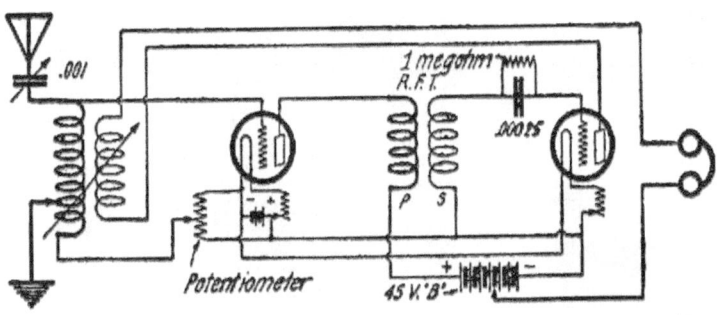

Fig. 54

Fig. 54. A receiver connected to a one stage radio frequency amplifier. The R. F. Transformer can be of any standard make. A variocoupler is again used with the rotor acting as a tickler.

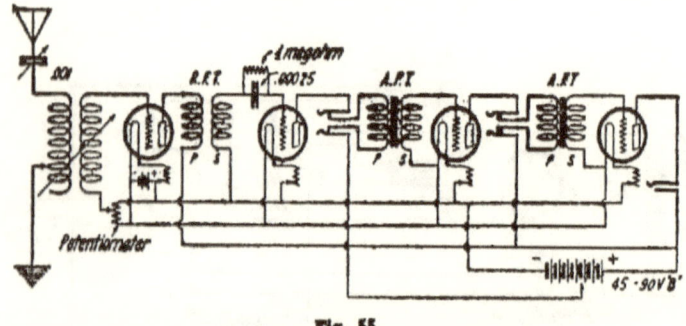

Fig. 55

Fig. 55. Here is a very good combination. A detector, one stage of transformer radio frequency amplification and two stages of audio frequency amplification. The radio frequency increases the range of the receiver while the audio frequency provides the volume.

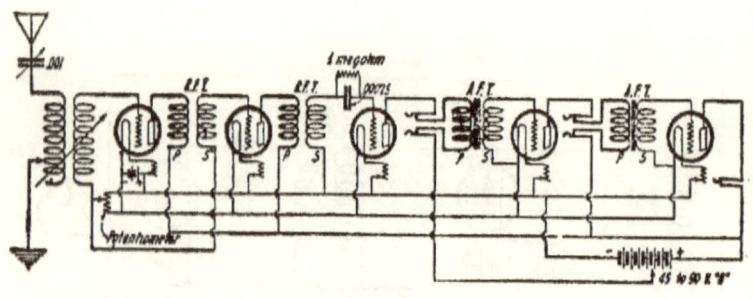

Fig. 56

Fig. 56. A receiver with the addition of two stages of radio frequency and two stages of audio frequency amplification. This arrangement is sensitive enough to be used with a loop aerial, although better results will be had by using it with an outside aerial as shown.

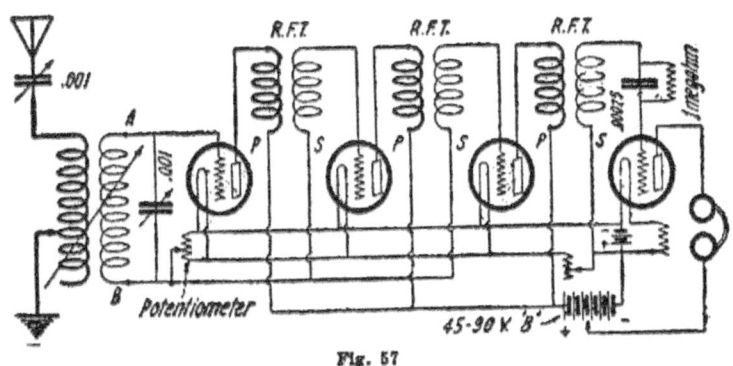

Fig. 57

Fig. 57. A two circuit receiver with three stages of radio frequency amplification. Such an arrangement is very sensitive indeed, and will pick up the long distance stations. Three standard radio frequency transformers are used as shown.

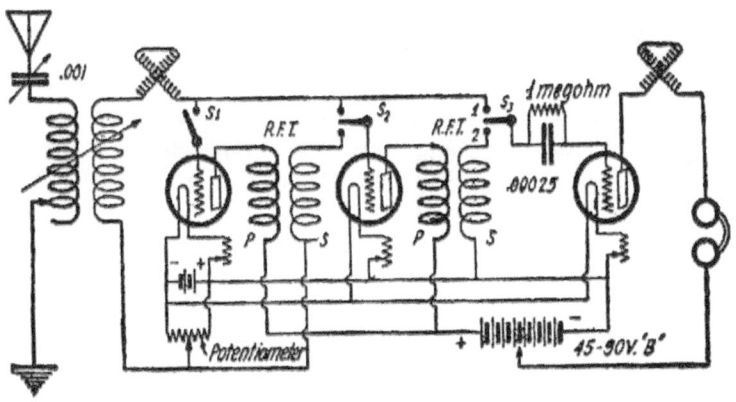

Fig. 58

Fig. 58. Here is a switching arrangement for cutting in or out the successive stages of radio frequency amplification which is often desirable. By placing S-3 on 1 only the detector is in use. By placing S-3 on 2 and S-2 up, the detector and one stage are in use. With S-3 on 2, S-2 down, and S-1 closed the detector and two stages are brought into operation.

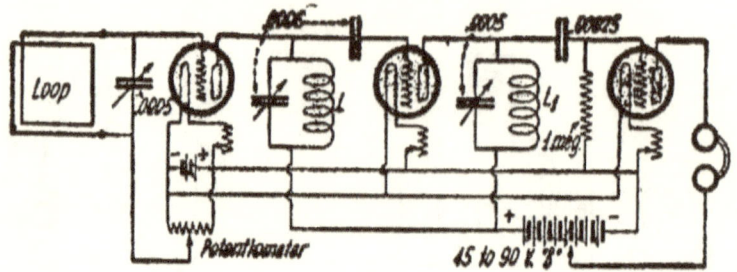

Fig. 59

Fig. 59. We have here a loop receiver with two stages of tuned radio frequency amplification. This type of amplifier is very efficient. The coils L and L-1 will have the same values and may be Honeycomb coils. However, 35 turns of wire on a 3 in. tube is the approximate value for broadcast wave lengths.

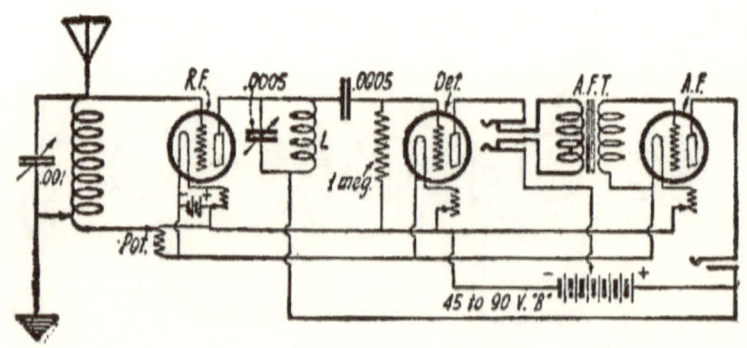

Fig. 60

Fig. 60. With this combination it is possible to cover a considerable range and obtain fair volume. It consists of one stage of tuned impedance radio frequency amplification and one stage of audio frequency amplification, in conjunction with a single circuit receiver. Coil L is approximately 35 turns of wire on a 3 in. tube, or a Honeycomb coil of a similar size.

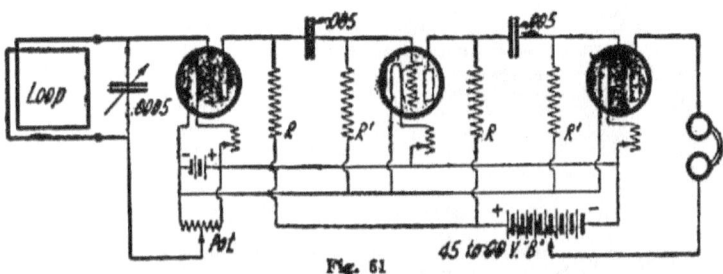

Fig. 61

Fig. 61. A detector and two stage resistance coupled radio frequency amplifier. R and R-1 are non-inductive resistances of 70,000 ohms and 1 megohm respectively. A loop aerial is shown since the outfit is sensitive, and capable of long distance reception.

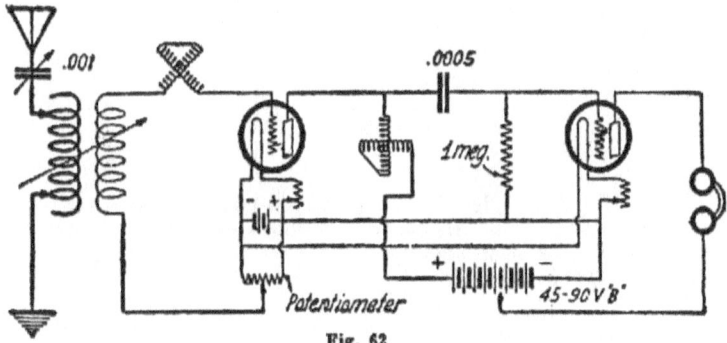

Fig. 62

Fig. 62. A two circuit receiver with one stage of tuned impedance radio frequency amplification. A variometer is employed as the impedance in this case, instead of a fixed coil and variable condenser. Such a set gives very good results and is easy to handle.

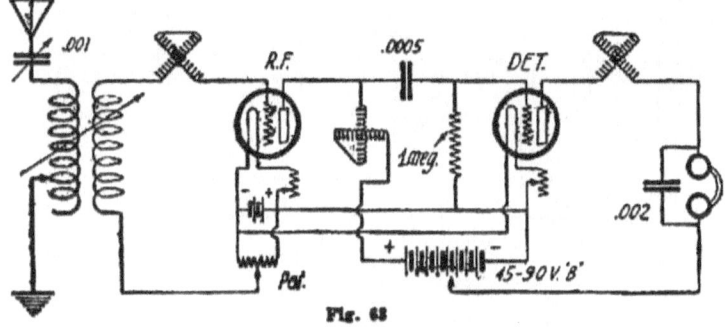

Fig. 63

Fig. 63. A circuit similar to that of Fig. 62 except that a third variometer is connected in the plate circuit of the last tube to produce regeneration. The variometers may be of any standard make.

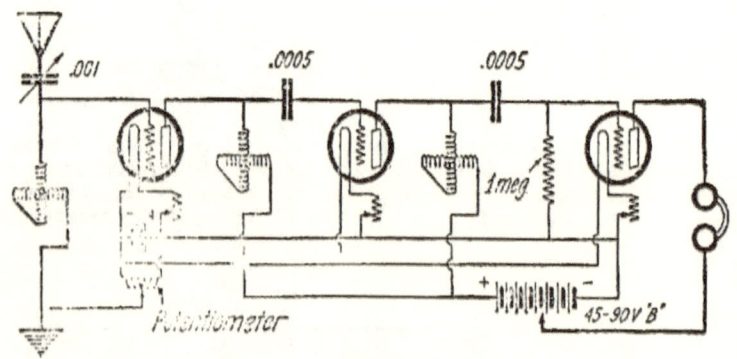

Fig. 64

Fig. 64. A simple single circuit receiver with two stages of tuned impedance radio frequency amplification using variometers throughout. For best operation the variometers should be from 6 to 8 in. apart.

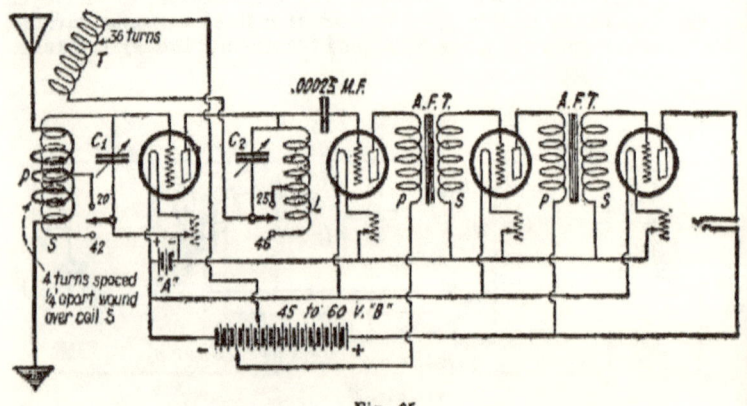

Fig. 65

Fig. 65. Here is the well-known Superdyne circuit. It employs, in conjunction with its single stage of tuned radio frequency amplification, a controlling or stabilizing factor referred to as "negative regeneration." The Superdyne compares favorably with the standard Super-Heterodyne when in the hands of an experienced operator. It is important that the two-stage audio frequency amplifier be used, as the circuit will not function properly without it. The coils are wound with No. 22 D.S.C. copper wire.

NEUTRODYNE RECEIVERS

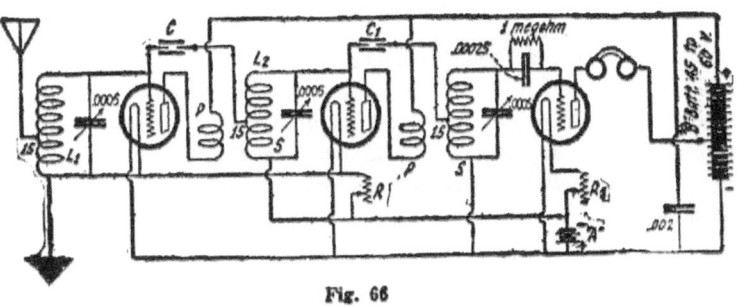

Fig. 66

Fig. 66. The circuit of the famous Neutrodyne receiver. This combination employs two stages of radio frequency amplification. One of its main advantages is the absence of all squealing noises since it is so wired that the radio frequency current cannot feed back through tubes A and B and produce oscillation. This is accomplished by condensers C and C-1 which are of a very low capacity (about one-microfarad). All coils are wound on three-inch tubing and with No. 26 D.C.C. wire. L consists of 60 turns with a tap-off at the 15th turn. L-2 and L-3 are identical to L, but each has a primary coil of 15 turns wound in the same direction and over the coil from the starting point to the tap. The Neutrodyne is very sensitive and selective and is a good long-distance receiver.

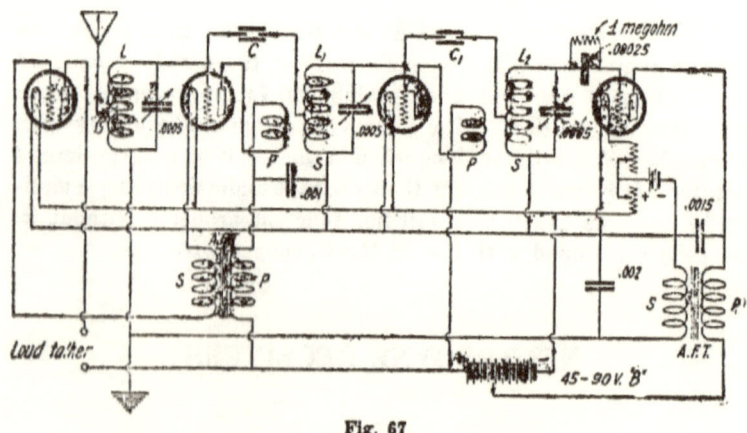

Fig. 67

Fig. 67. A four tube Neutrodyne circuit having two stages of radio frequency amplification and two stages of audio frequency amplification, thus adding volume to distance. The constants of the circuit are the same as those mentioned in conjunction with Fig. 66. The audio frequency transformers may be of any standard make.

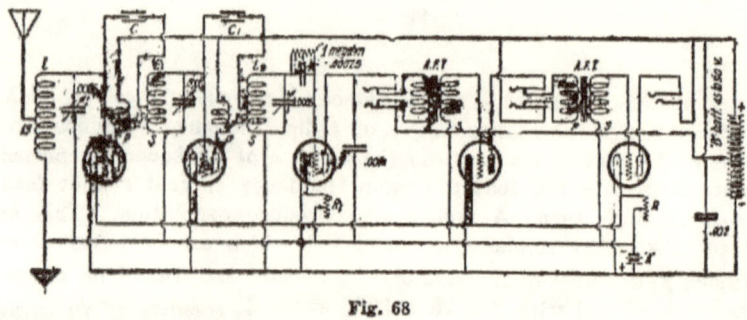

Fig. 68

Fig. 68. A five tube Neutrodyne circuit. This is the same as that of Fig. 66 but with the addition of a two-stage audio frequency amplifier. It is more reliable than the four tube circuit of Fig. 67, being less critical in adjustment.

REFLEX CIRCUITS

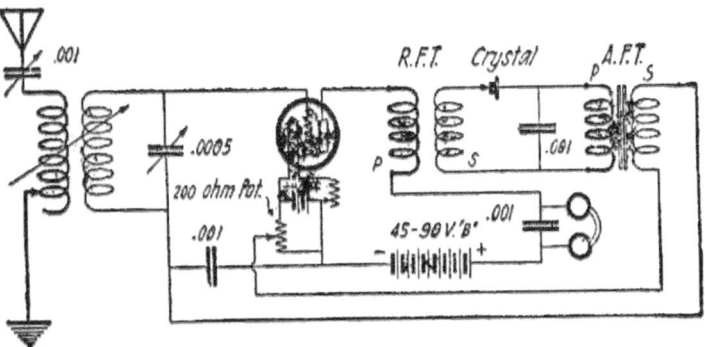

Fig. 69

Fig. 69. A one tube reflex circuit wherein the same tube is used for both a radio and audio frequency amplifier. This is of the one radio frequency one audio frequency type. A crystal detector is employed for rectification of the incoming signals Standard radio and audio frequency transformers are employed.

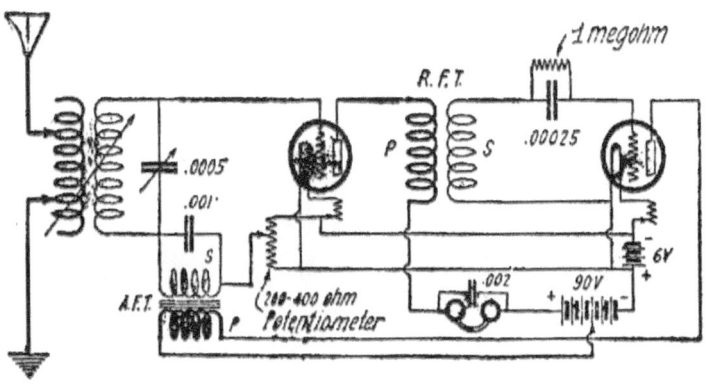

Fig. 70

Fig. 70. A two tube reflex set with one stage of R.F. amplification and one stage of A.F. amplification. The second tube acts as the detector.

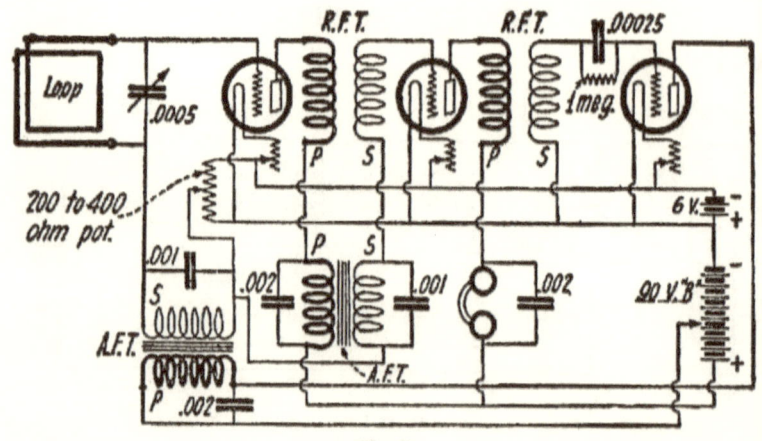

Fig. 71

Fig. 71. A three tube reflex receiver with two stages of R.F. amplification and two stages of A.F. amplification. The third tube acts as the detector. This set is very effective when used with a loop aerial.

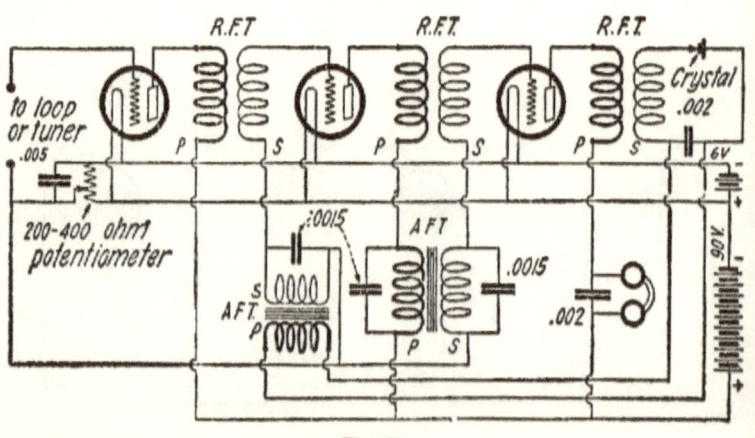

Fig. 72

Fig. 72. A three tube reflex set wherein it is possible to have three stages of R.F. amplification and two stages of A.F. amplification. Rectification is accomplished by a crystal detector. Either a loop or outside antenna can be used.

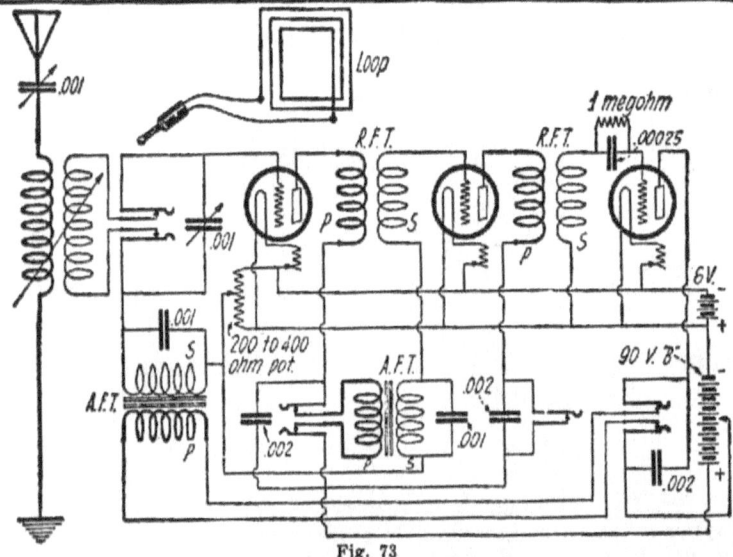

Fig. 73

Fig. 73. Here is a three tube reflex set with two stages of R.F. amplification and two stages of A.F. amplification which are controlled by jacks. One jack is used for plugging in a loop when desired while the others provide means for plugging the headphones or loud speaker in the successive stages of amplification.

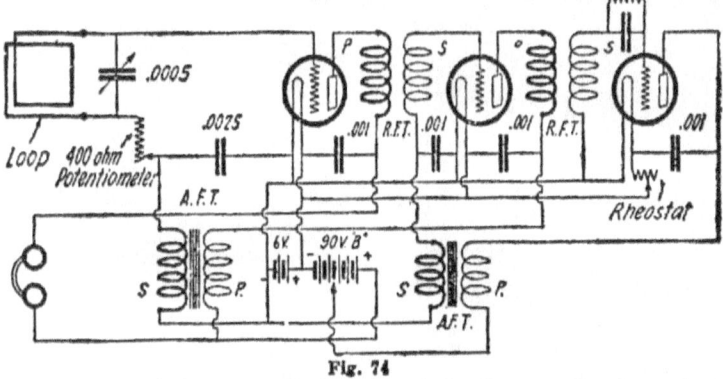

Fig. 74

Fig. 74. This circuit is known as the Inverse Duplex. It is similar to those of the reflex type except that the work is evenly distributed between the three tubes so that none of them are overloaded. The paths for the radio frequency currents are made shorter than in the usual reflex set. The Inverse Duplex receiver shown here has two stages of R.F. amplification and two stages of A.F. amplification. Standard transformers may be used.

SUPER-REGENERATIVE CIRCUITS

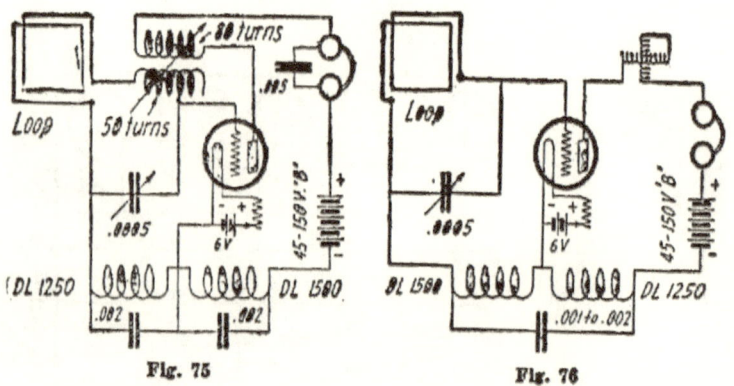

Fig. 75   Fig. 76

Fig. 75. A simple single tube super-regenerative receiver employing a loop aerial, a variable condenser and a variocoupler for tuning. The rotor of the variocoupler is used as a tickler coil. It should be rewound with from 80 to 90 turns of wire.

Fig. 76. A simple form of one tube super-regenerative receiver. All of the tuning is accomplished by the variable condenser across the loop, while regeneration is produced by a variometer. The last mentioned can be of any standard make.

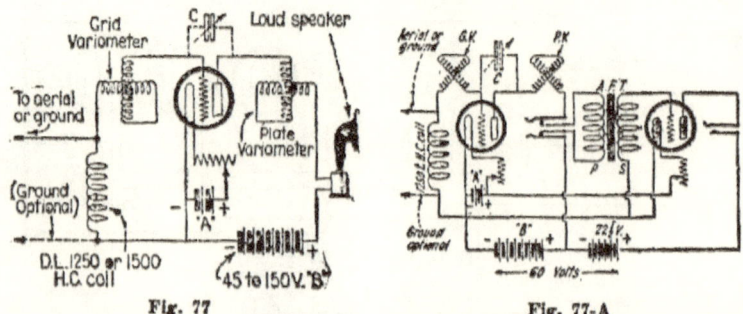

Fig. 77   Fig. 77-A

Fig. 77. This, the Autoplex circuit, is a simplified form of super-regenerator. Under proper guidance it proves a good long-distance receiver and usually provides enough energy for loud speaker operation. The vernier variable condenser C is not a necessity, but helps to boost the wavelength range of the set and eliminate circuit noises.

Fig. 77-A. This shows how a one-stage audio frequency amplifier is added to the Autoplex circuit. The "B" battery is common to both tubes.

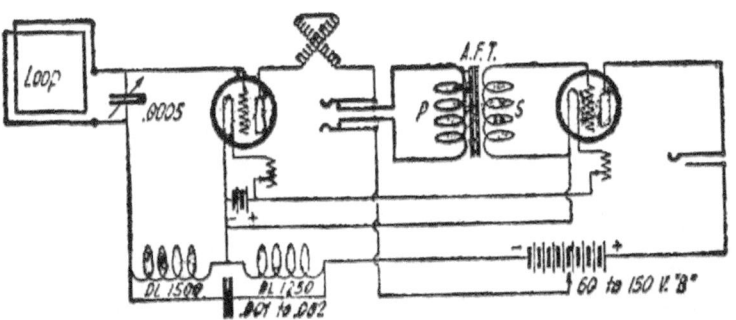

Fig. 78

Fig. 78. The same hook-up as that of Fig. 76 with the addition of a one stage amplifier. By using a .001 M.F. fixed condenser across the large Honeycomb coils, the variation frequency (which is always evident while receiving) is raised to a very high pitch. The audio frequency amplifying transformer fails to amplify it much due to its inefficiency at high frequencies. It is therefore not bothersome. This circuit should only be used with a loud speaker as the volume is too great for a pair of headphones to handle.

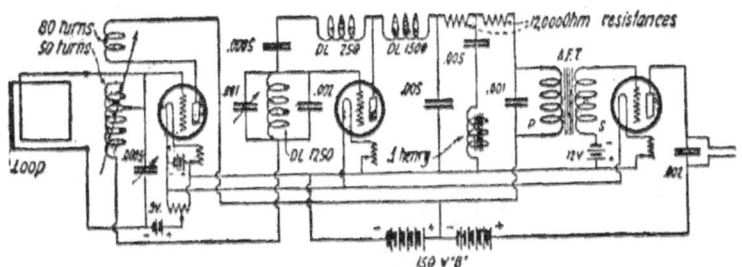

Fig. 79

Fig. 79. This is the original Armstrong three tube super-regenerative receiver, the first tube being the regenerator and detector, the second tube the oscillator and the third an audio frequency amplifier. The tickler coil T is the rotor of a variocoupler rewound with 80 to 90 turns. A filter system consisting of resistances, condensers and a choke coil are introduced so that the variation frequency is choked back and kept out of the amplifier circuit.

41

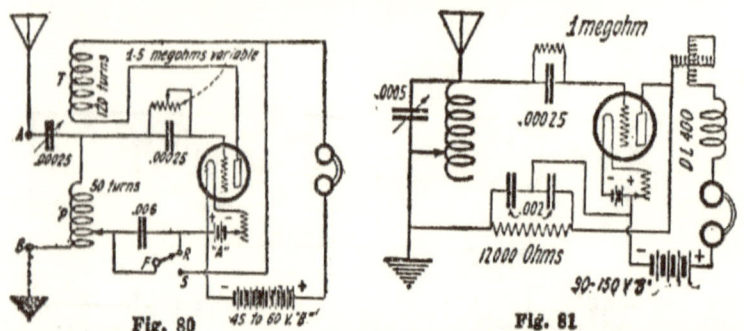

Fig. 80   Fig. 81

Fig. 80. This is the improved Flewelling super-regenerative circuit. Coils P and T are those of a variocoupler rewound with the correct number of turns. Switch F changes the circuit from regenerative to super-regenerative at will.

Fig. 81. The Bishop Ultra Regenerator, named after its originator. It is a simplified form of the Flewelling and is superior in its operation. A variometer is employed for obtaining regeneration. The Honeycomb coil D.L. 400 acts as a radio frequency choke.

SUPER HETERODYNE RECEIVERS

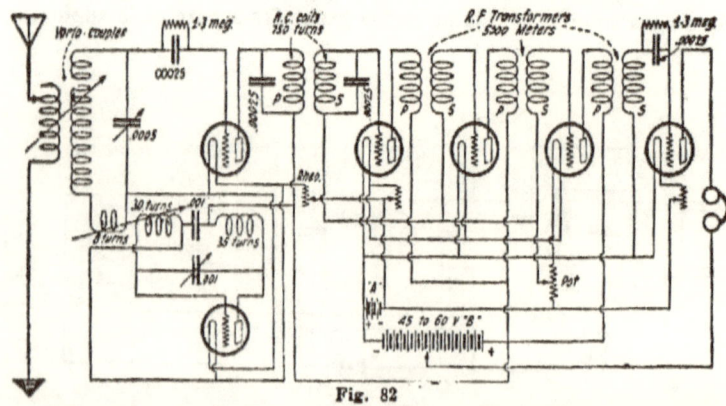

Fig. 82

Fig. 82. The well-known Armstrong Super-Heterodyne circuit. This is mainly an improved form of radio frequency amplifier, but has other advantages, such as simplicity of control, and selectivity. The radio frequency transformers in this circuit are designed to amplify at a wavelength of approximately 5,000 meters and are not of the type employed in the usual radio frequency amplifier. The constants of the apparatus are given in the diagram.

42

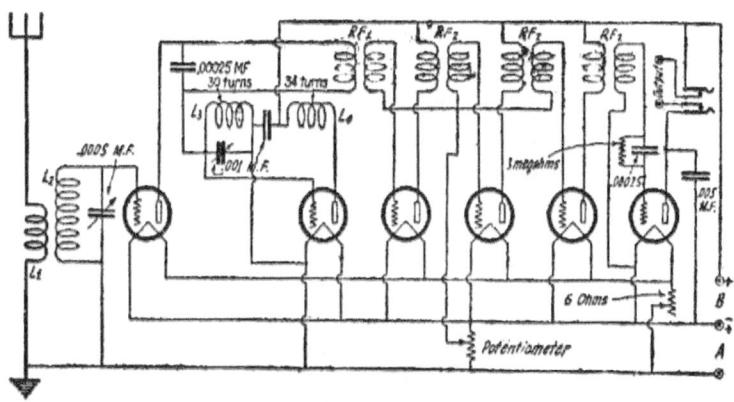

Fig. 83

Fig. 83. This is the circuit of the Ultradyne, an improved form of Super-Heterodyne, and is, without a doubt the most sensitive circuit in existance at the present time. It employs what has been called the "modulation system" of rectification in place of the usual "frequency change" in a standard Super-Heterodyne.

MISCELLANEOUS HOOK-UPS

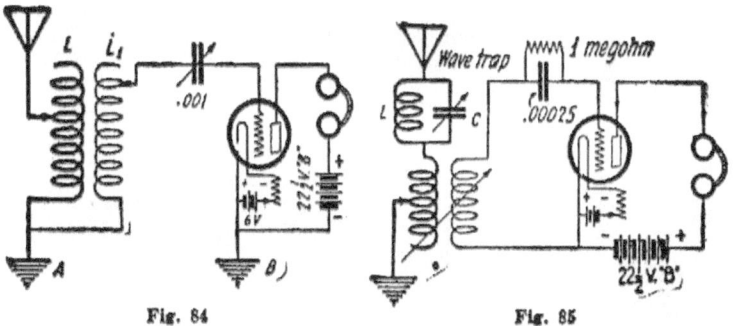

Fig. 84   Fig. 85

Fig. 84. A uni-polar vacuum tube receiving set. This can be used to advantage where there is a considerable amount of local interference, especially that produced by arc lights and alternating current power lines. This circuit is suitable for the reception of both damped and undamped wave signals and is quite sensitive. The coils L and L-1 can be the primary and secondary of a loose coupler.

Fig. 85. A circuit employing a wave trap for the elimination of an interfering station. When the wave trap is tuned to the same wave length as the interfering station, its signals are choked back, all signals of other wave lengths, however, can pass the trap freely. For broadcast reception, L can be a D.L. 35 Honeycomb coil and the variable condenser C be of .001 M.F. capacity. They should be shielded from the other apparatus.

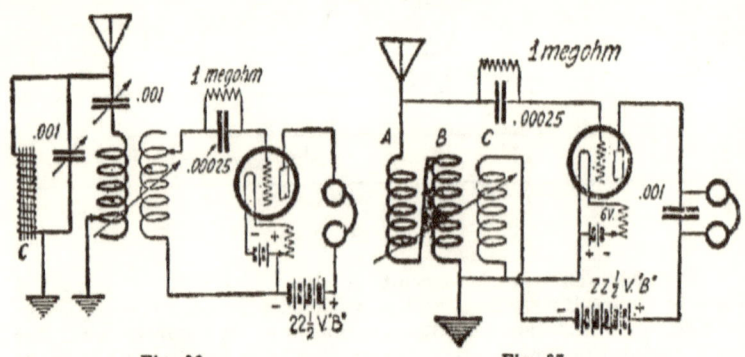

Fig. 86   Fig. 87

Fig. 86. A circuit for the elimination of bothersome A.C. hum and other interference of the same nature. C is an iron core choke shunted by a .001 M.F. variable condenser. The correct value for C can only be determined by experiment, as its size depends on the frequency of the interference.

Fig. 87. A single circuit regenerative receiver with but two controls; tuning and regeneration being accomplished by the movement of coils A and C. Coils A and B should have about 60 turns each and coil C about 35 turns. They can be wound in a pancake form or in spider-web fashion, and placed in a suitable mounting.

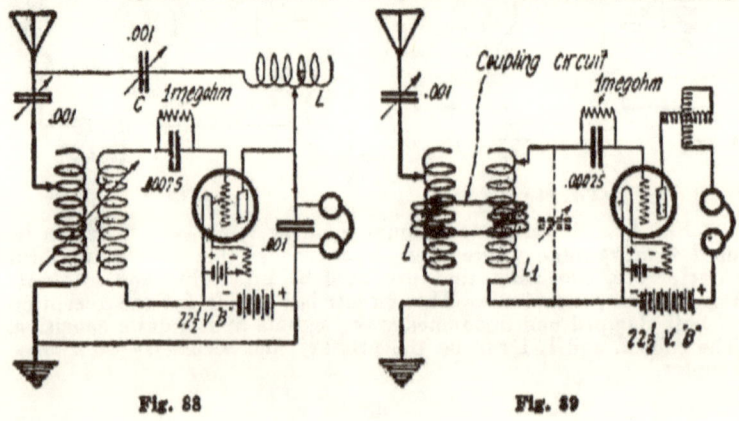

Fig. 88   Fig. 89

Fig. 88. Another arrangement for eliminating interfering stations. When the circuit C-L is tuned to the same wave length as the interfering station, its signals will vanish. L is a tapped coil of about 50 turns.

Fig. 89. Here is a circuit with a fixed coupling, developed by the Marconi Company some years ago. It is very selective and has the advantage of few controls. The coupling circuit consists of a continuous piece of insulated wire with about three turns wound around the outside of each tuning coil. The tuning coils L and L-1 should be from 4 to 6 in. apart or, at right angles.

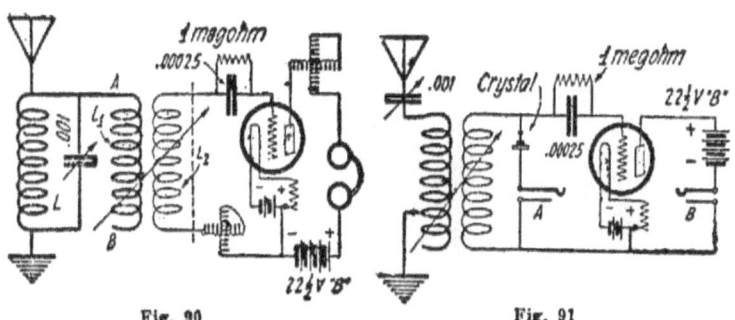

Fig. 90   Fig. 91

Fig. 90. An adaption of the Pearce circuit, developed by him some years ago in an attempt to cut out interference. It is very selective and very easy to tune. Coil L consists of approximately 23 turns on a 3 in. tube. L-1 and L-2 are the coils of a vario-coupler. If good results are not obtained with the aerial connected to point A on coil L-1, it should be shifted to point B. L should NOT be in inductive relation to coils L-1 and L-2.

Fig. 91. A combination vacuum tube and crystal receiver. When the headphones are plugged into jack A, the crystal is in use. When plugged into jack B the vacuum tube can be operated. The tube should not be lit when the crystal is being used. Open circuit jacks should be employed.

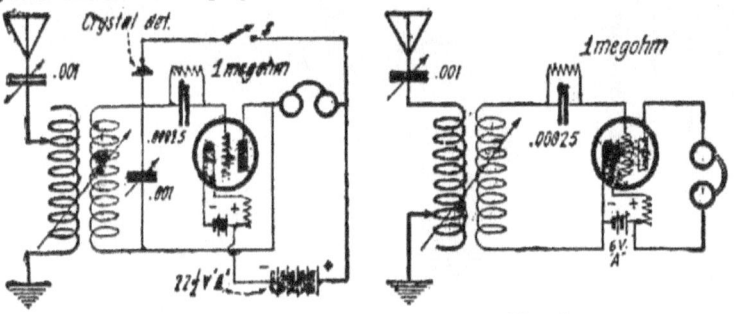

Fig. 92   Fig. 93

Fig. 92. Another similar arrangement wherein the crystal detector may be used by closing the switch S and putting out the tube.

Fig. 93. Here is a vacuum tube hook-up without a "B" battery. It operates best when a soft tube is used. Such an arrangement comes in handy when the "B" battery goes dead. Note that one side of the phones connects to the positive terminal of the "A" battery.

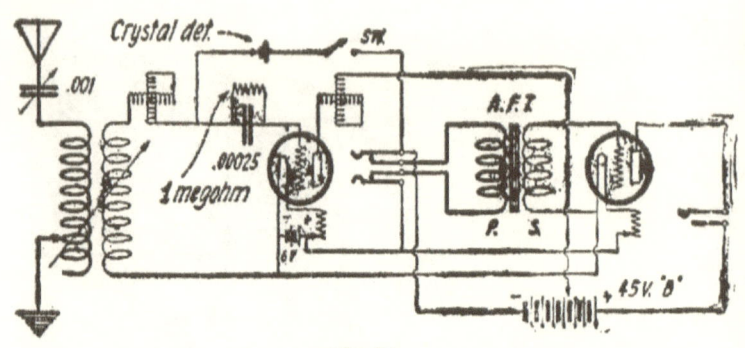

Fig. 93

Fig. 94. A three circuit regenerative receiver in which there is the preference of the vacuum tube detector and the one stage audio frequency amplifier, or the crystal detector and the amplifier. The crystal is brought into operation by closing switch SW and extinguishing the tube.

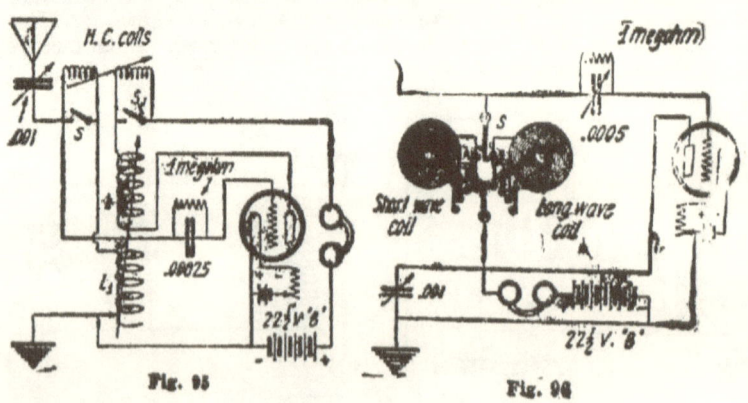

Fig. 95   Fig. 96

Fig. 95. A combination long and short wave receiver of the regenerative type. Long waves can be received when switches S and S-1 are open. L and L-1 are the coils of a variocoupler. The Honeycomb coils should be of such a value to cover the long wave lengths desired to be received.

Fig. 96. Another form of short and long wave receiver which is also very sensitive. The diagram is self-explanatory.

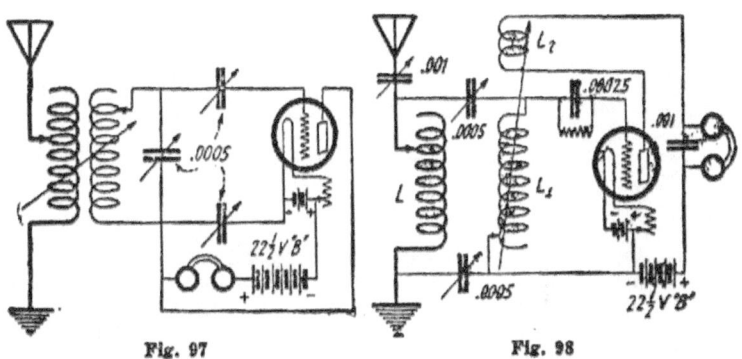

Fig. 97   Fig. 98

Fig. 97. A capacity coupled receiver using a variocoupler and three variable condensers. This arrangement is very selective indeed.

Fig. 98. Another type of capacity coupled receiver with the addition of a tickler coil for regeneration. L can well be a one slide tuning coil while L-1 and L-2 are the coils of a variocoupler. Most of the tuning is accomplished by the variable condensers.

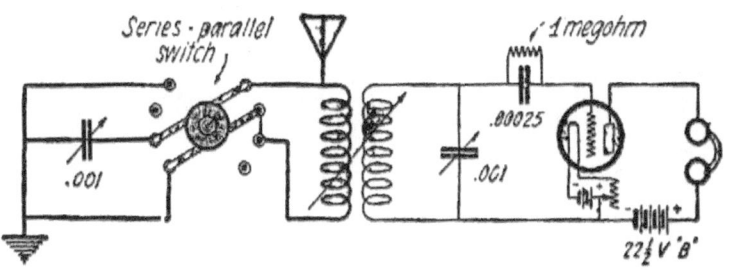

Fig. 99

Fig. 99. A circuit with the addition of a switch for connecting the variable condenser in series, out of the circuit, or in parallel with the primary coil of the coupler.

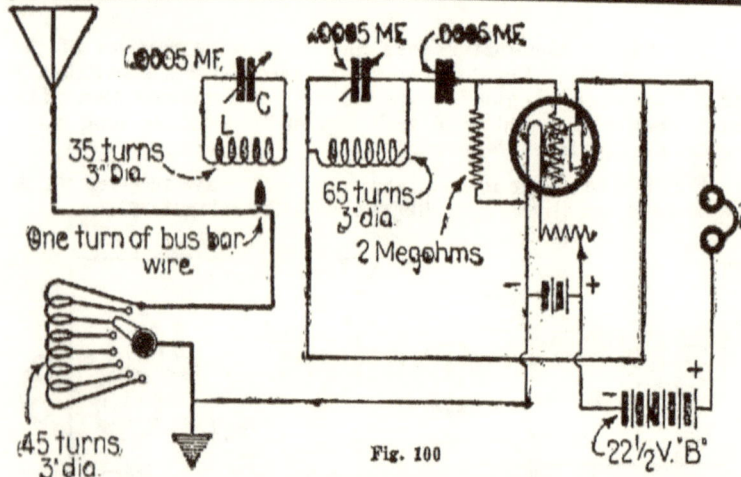

Fig. 100

Fig. 100. The hook-up for the Cockaday four circuit tuner is shown above. This in reality is the Ultra-Audion circuit used in conjunction with a "sensitizing circuit", L-C. By the addition of this fourth circuit, perfect regeneration control is made possible. The primary and secondary circuits are coupled by means of one turn of wire. The primary of 45 turns is double bank wound and is placed in non-inductive relation to the secondary and sensitizing coils. The antenna circuit is semi-aperiodic. Long distance reception is claimed for this receiver.

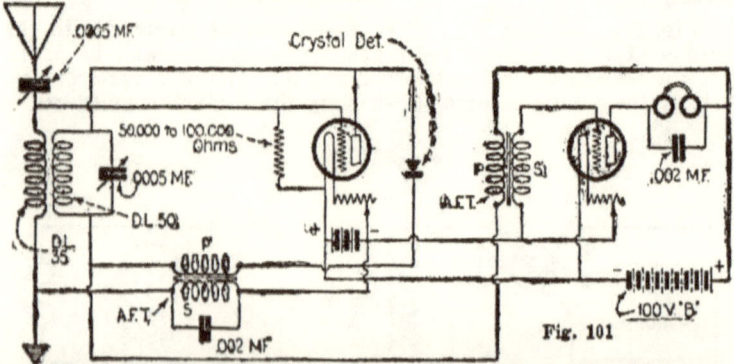

Fig. 101

Fig. 101. The ST-100 circuit has gained considerable popularity in England and is fast becoming a favorite here. It is a form of Reflex circuit, but so connected as to produce audio frequency regeneration. A loud-speaker can be used with this circuit when listening to nearby stations, although phones are preferable when receiving over long distances. It is claimed that distortionless amplification is obtained with this circuit.

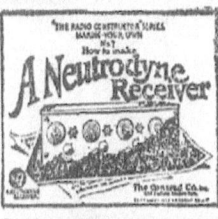

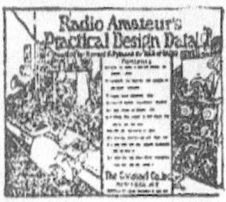